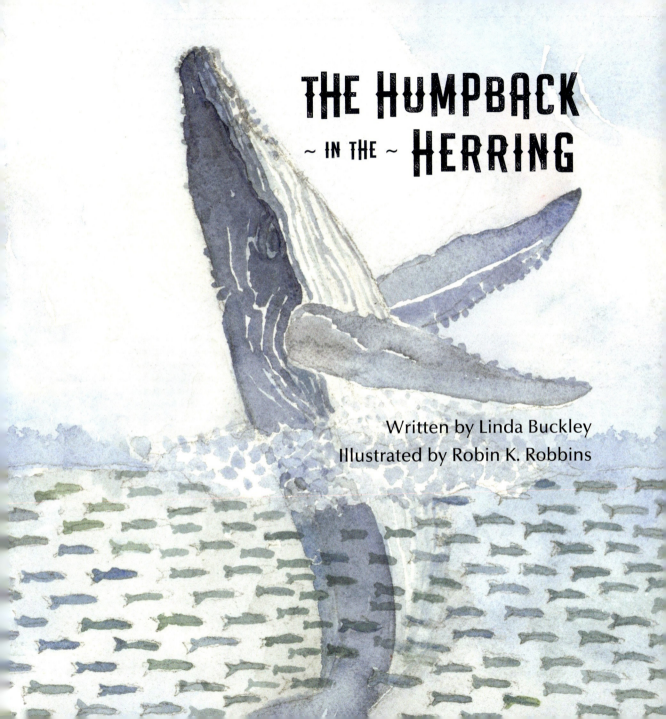

The Humpback in the Herring
A children's book on deep ecology

Text: Linda Buckley

Watercolor Illustrations: Robin K. Robbins

Whale Fact Consultant: Dr. Fred Sharpe

Layout & Design: Liorah Wichser at Blue Tiger Studio

Copyright © 2022 Linda Buckley
All rights reserved

Humpback drawing by Luke Buckley (grandson of the author)
Whale fluke drawings by Elias Buckley (grandson of the author)

Designed in the USA
Printed in Canada by Friesens
Second printing: 2024

Alaskasong, Inc. Publisher
Juneau, Alaska
lindagramma@gmail.com

ISBN 978-0-578-98161-1

Dedicated to my son, Jim.
He has reawakened my sense of wonder and curiosity
since he was old enough to turn over a rock.

The Humpback
~ in the ~ Herring

Can you see the humpback in the herring?

No?

Can you see the sun in the herring?

Yes.

Can you see the rain in the herring?

Yes.

Can you see the ocean in the herring?

Yes.

What happens when the whale eats a ton of herring?

He gets so full he can barely swim. It's time for a nap.

He floats on the surface while sleeping.

When he wakes up, he has a tummy ache. He poops and feels much better. The poop is called "feces."

The poop, I mean feces, looks like a big yellow cloud. It is filled with something called nitrogen. All life in the ocean needs nitrogen, even the herring.

North Pacific humpback whales feed in Alaska in the summer.
Most swim to Hawaii in the winter to have their babies.
(Some travel to Mexico and even all the way to Japan.)

It is nearly 3000 miles from Alaska to Hawaii.
By swimming day and night they can make the journey in one month (if they don't lollygag).

Like most songbirds, only the males sing. Scientists are still discovering the role of song.

Did you know that whales can sing? The songs are how the males find a female.

The female whales become pregnant.

It will take almost a year until their babies are born.

Baby whales are called "calves."

The pregnant female swims back to Alaska to feed and get strong. In the fall, she makes the long journey back to Hawaii to have her baby.

The bond between the mama whale and the baby calf is very strong. The new calf stays very close to her mother.

When only a month or two old, they swim together back to Alaska. The calf stays with mom for nearly a year and learns to feed, sing and avoid danger.

The cycle repeats year after year. The whales feed in Alaska then swim to Hawaii.

In Alaska, there are lots of herring.

The herring are filled with nutrients from the whale poop. Oops, I mean feces!

Now can you see the humpback in the herring?

Keep looking...

Remember the whale poop? Where did it go?

The rain, the sun, the ocean, and even the humpback is in the herring. Everything is in everything.

It's the circle of life spinning around in the vast ocean.

THE END

NOW,

Follow the flukes and learn some

"whale facts" that might surprise you.

WHALE FACTS:

How much food does a humpback whale eat per day?

An average sized humpback will eat four to five thousand pounds of food each day in the summer months.

What do humpback whales eat besides herring?

They eat squid, krill, salmon and other fish.

How much does an adult humpback weigh?

An adult humpback whale weighs over 50,000 pounds! Wow. They are the size of a school bus.

What is baleen?

Instead of teeth, baleen hangs down from the roof of the mouth. The whale gulps prey and seawater, then filters the water back out through the baleen.

How do scientists identify humpback whales?

Each whale's tail is different. Their tail is like our fingerprint. Scientists take photographs of the whale's tail and then they can track them. They can learn where they are eating, how many calves they have, if they are injured and more.

PHOTO CREDIT: ERIC AUSTIN YEE

PHOTO CREDIT: MICHAEL SMITH

What is bubble net feeding?

Humpbacks often hunt in teams. One whale swims in a circle below the surface while blowing air out of their blowhole. The rising air forms a "bubble net" to trap the herring. Other whales give a loud sound to confuse the herring and chase them into the bubble net. The fish feel trapped in the bubble net as the whales torpedo upwards engulfing the fish in their giant mouths.

What are some of the acrobatics of humpback whales?

Whales jump out of the water making a big splash. This is called "breaching." They also make a loud noise lobing their tails like beating a drum. To look around they do something called "spy hopping."

How far does the humpback's song travel?

Their sounds may travel several hundred miles. Unbelievable, right?

How do large ships affect humpback whales?

If they don't pay attention, large ships and small fast moving boats, can wound or kill whales by running into them. The noise of vessels makes it harder for the whales to find each other, hear their prey, and avoid predators.

Are humpbacks dangerous to humans?

No. It is important to give them space. The Marine Mammal Protection has rules stating we should not get closer than 100 yards. That is the same as one football field. The Coast Guard helps enforce these important rules.

How are humpback whales affected by climate change?

There are many factors affecting humpbacks. One of the greatest is ocean warming which leads to ocean acidification. This can destroy phytoplankton at the bottom of the food chain.

What are phytoplankton?

They are tiny plants in the ocean that form the base of marine food webs.

Some good websites to learn more about humpback whales:

happywhale.com

alaskawhalefoundation.org

alaskatrust.org

nps.gov/glba

About the Author

Linda Buckley has lived in Alaska for over 50 years. She taught music, art, outdoor education, and Alaska culture classes. She also served as the librarian in a small rural school in Southeast Alaska. Her songs have been recorded on four CD's. She published her first children's book, "The Bear in the Blueberry" in 2019. "The Humpback in the Herring" is the sequel. Linda is an active environmentalist and hopes that these books will inspire young people to care for the earth as they understand how everything is connected to everything.

About the Illustrator

Robin Kinney Robbins lives in Seattle, Washington. Robin taught in the Peace Corps and in Washington Public Schools for 38 years. She taught art alongside bilingual education. Robin travels with her paints, not with a camera. She has painted in Peru, Ecuador, Mexico, Cambodia, Vietnam, Laos, and Europe. She has exhibited her work in galleries in Seattle and in her second home in Chautauqua, New York. She illustrated her first book, "The Bear in the Blueberry" and this is her second children's book.